ANIMAL

Coloring Books for Kids Ages 8-12

Nick Marshall

C000070226

This Coloring Book Belongs to

_____

Copyright © 2019 by Nick Marshall All rights reserved.

No part of this book may be reproduced in any form or by any electronic or mechanical means, including information storage and retrieval systems, without written permission from the author, except

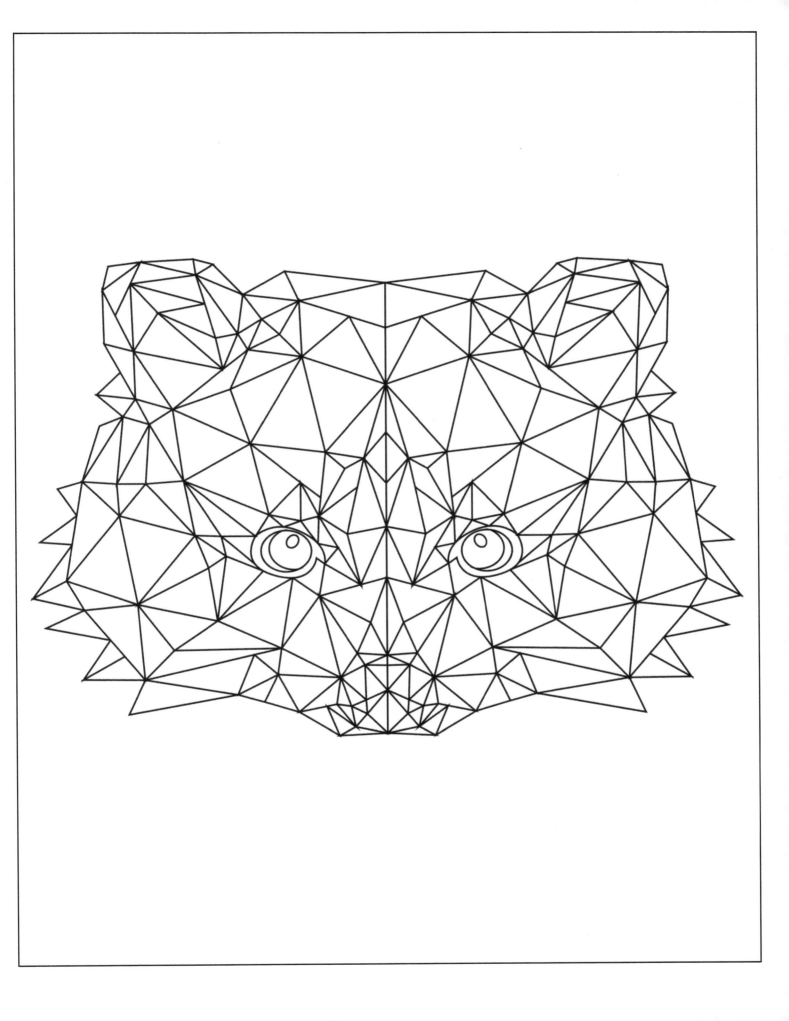

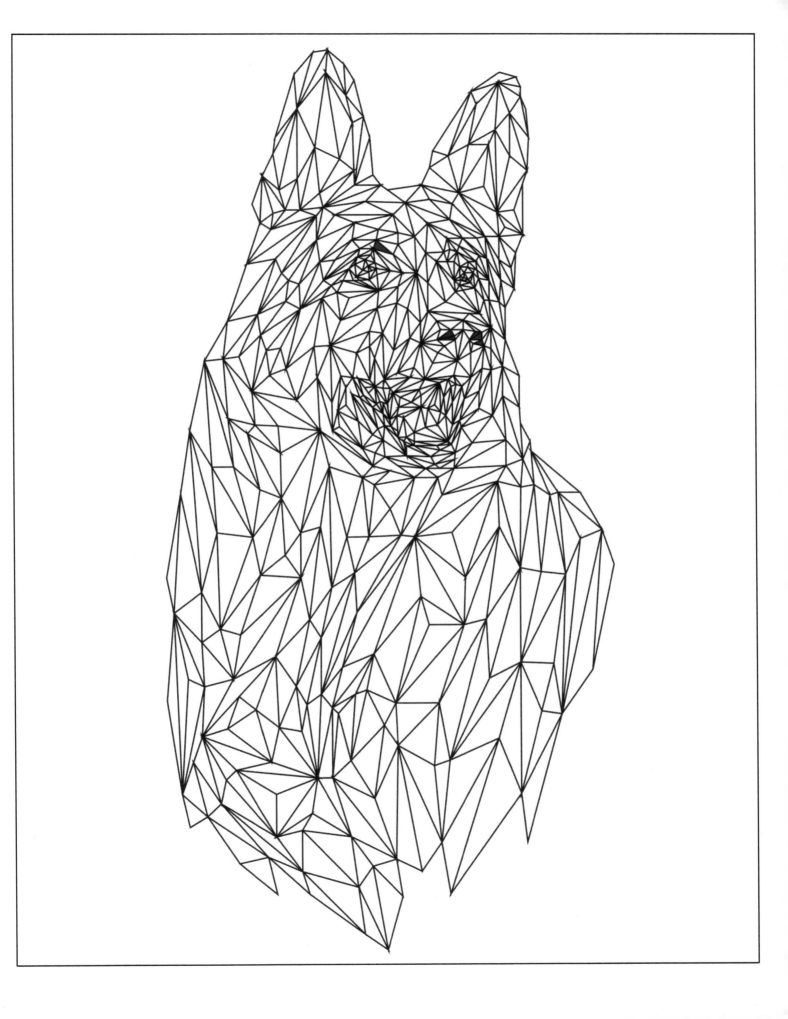

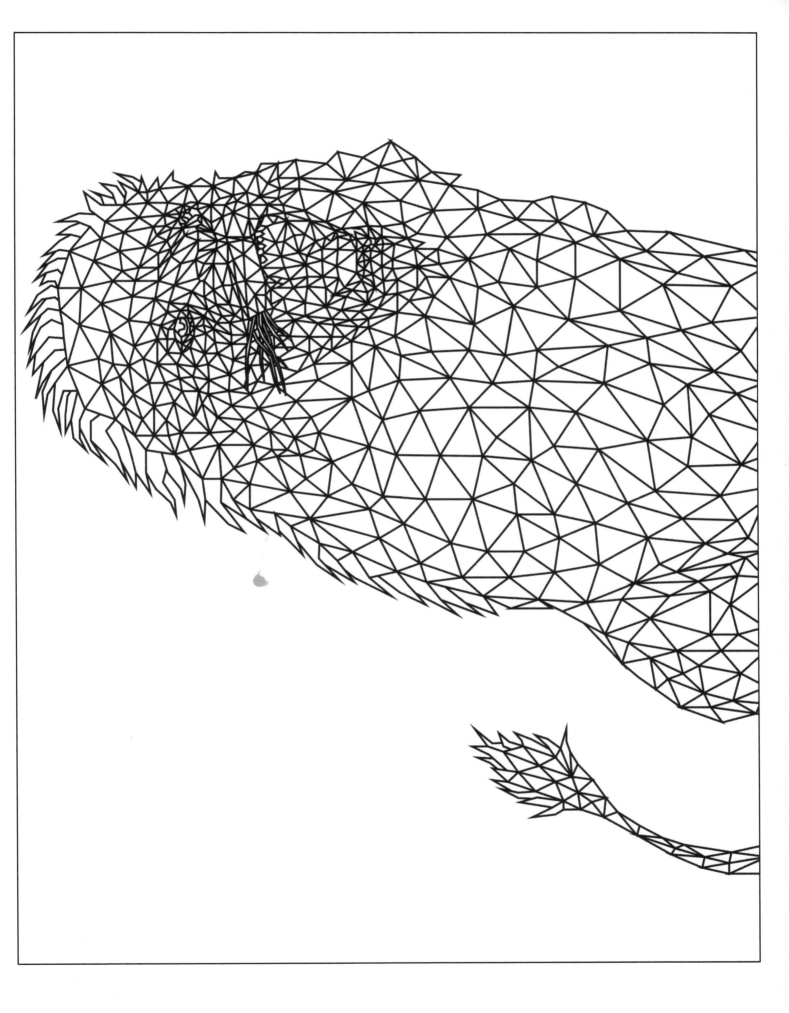

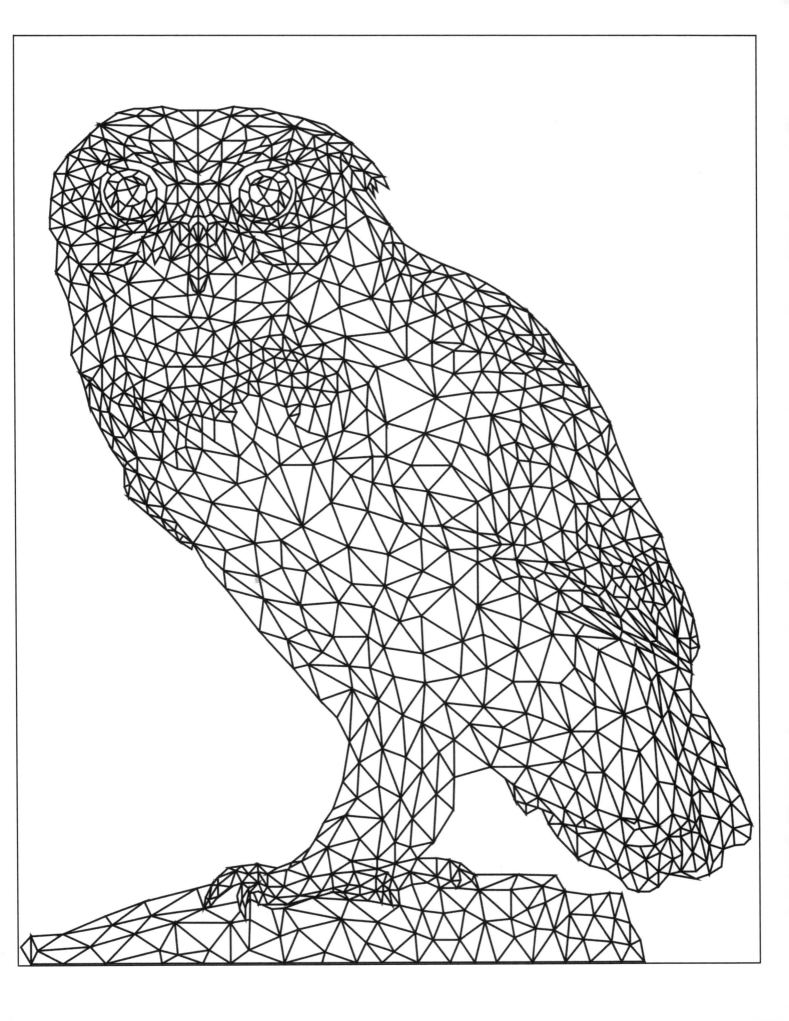

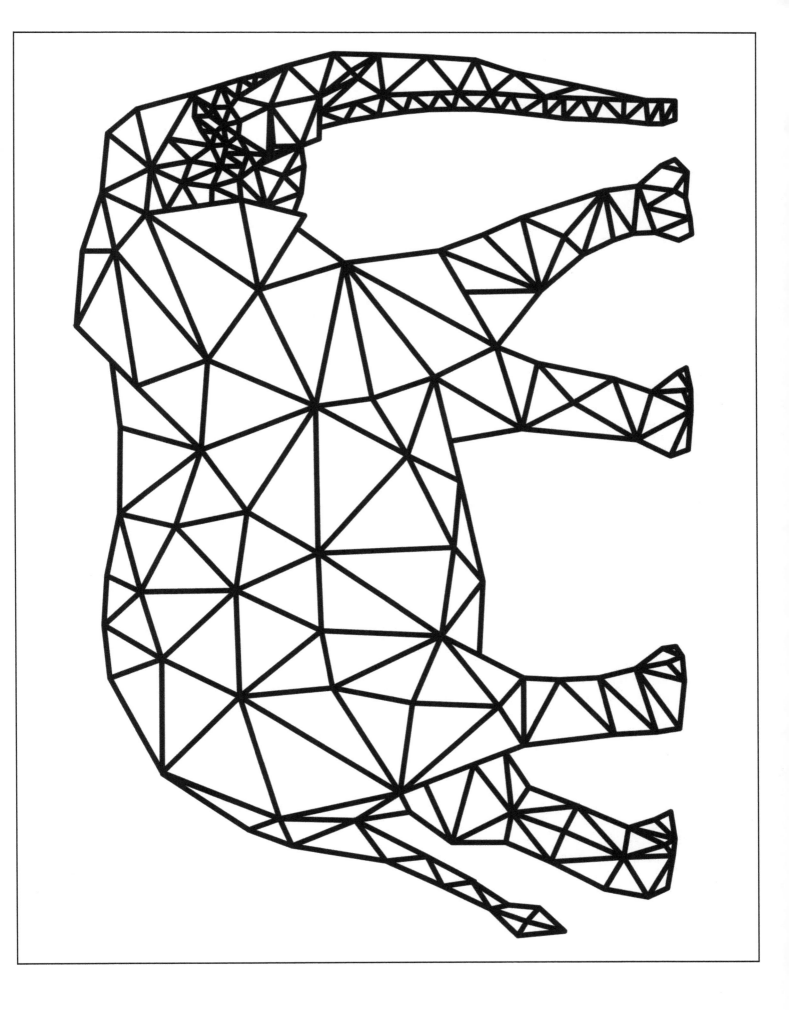

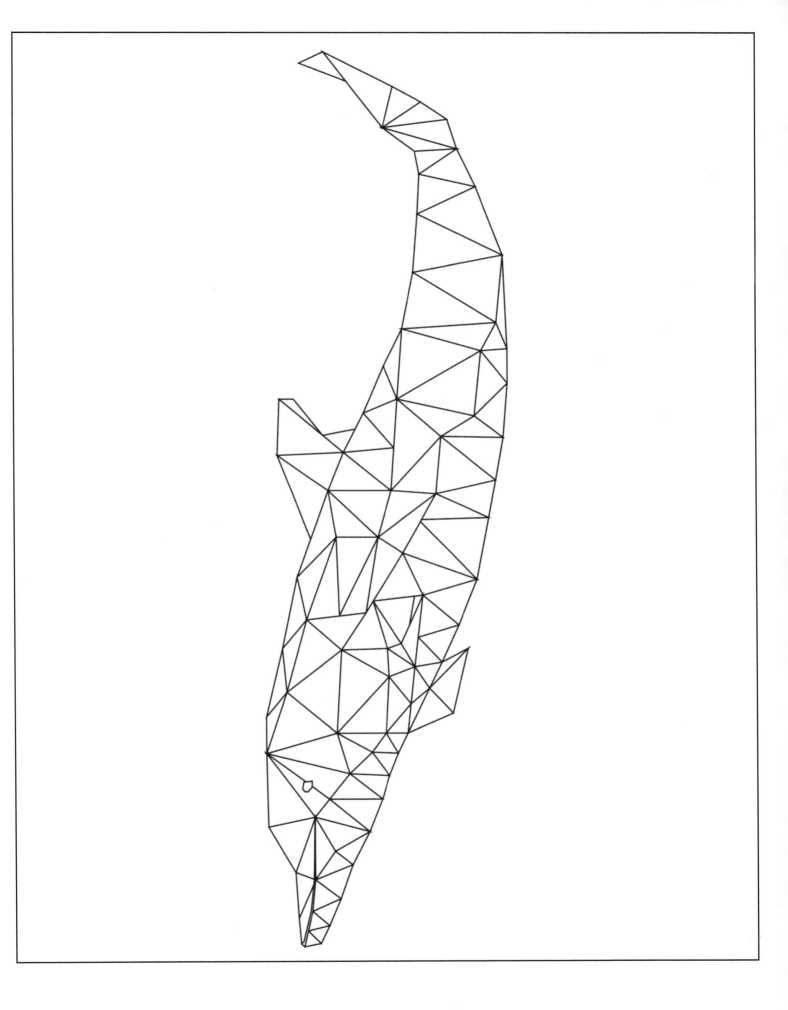

Lightning Source UK Ltd.
Milton Keynes UK
UKHW032115271121
394668UK00003B/312